MUSHROOM FINDER

Identifying Fungi in North America

JACOB KALICHMAN
illustrated by ROO VANDEGRIFT

Nature Study
Guild Publishers
an imprint of AdventureKEEN

T0191125

ABOUT THE AUTHOR/ILLUSTRATOR

Jacob Kalichman is a field mycologist from California and Tennessee. While studying logic at Stanford University in 2010, he became devoted as an amateur to wild mushroom identification and photography. With an ongoing interest in identifying gilled mushrooms in difficult genera, he has also taken up digesting and sharing updates in mycological taxonomy. He compiled a compendium of generic names of agarics and Agaricales, published as a paper in 2020 and continually updated at www.agaric.us, where he also maintains suggested common names and associated guidelines for North American fungi. He wrote the current edition of the *National Audubon Society Mushrooms of North America*.

Roo Vandegrift, PhD, is a queer scientist and illustrator. He received his doctorate in mycology from the University of Oregon and is best known for his research in tropical ecology and the fungal family Xylariaceae. As a National Geographic Explorer, Roo coordinated a multi-disciplinary expedition to the Los Cedros Biological Reserve in Ecuador, and he is producing the forthcoming documentary film *Marrow of the Mountain*. He is currently the Plant Pathologist at the USDA Plant Inspection Station in San Diego.

Mushroom Finder
Copyright © 2024 by Jacob Kalichman (text) and Roo Vandegrift (illustrations)
Published by Nature Study Guild Publishers
An imprint of AdventureKEEN
naturestudy.com
Printed in China
ISBN 978-0-912550-41-1
Cataloging-in-Publication data is available from the Library of Congress.

HOW TO USE THIS BOOK

This is not a typical mushroom guide. A typical mushroom guide uses photographs and lengthy descriptions—as it must—to help you identify mushrooms to species (or genus) level. With a pocket-size format and line-art illustrations, this guide is aimed not at depth, but at breadth. Usually the identification it provides will be more general (perhaps much more general) than a species or a genus. If you use this book for all the most eye-catching mushrooms you find, you will eventually be disappointed to repeatedly end up on only a few of the most common categories. When you reach this point, slow your hiking pace, and pay attention to smaller, less conspicuous, or weirder growths—even if you're not sure they're fungal. This guide should, superficially, be close to comprehensive for your finds.

More specifically, this book should cover everything that meets all of these criteria in the continental United States and Canada:

- is or looks like a fungus

- is at least 5 mm tall, wide, or thick. Patches >5 mm wide of units <5 mm that are obviously separate do not count.

- has some tangible thickness in all dimensions. In other words, this book does not include lesions or discolorations on leaves (leaf spots), wood, or walls.

- is more or less fresh. All mushrooms will eventually become shriveled, tough, and crispy in a dry enough environment, and all mushrooms will eventually become

1

squishy and melty (rotten) in a wet enough environment. Old mushrooms are not worth trying to identify. It is not always obvious whether a mushroom is fresh or not, but your intuition is always useful and will improve with practice.

Read the glossary first. All terms defined there are used later in the book. Try leads in order. Most leads depend on the assumption that the earlier leads in the couplet did not apply to your find. There are exceptions to almost every rule about how fungi look, feel, and smell. There are intermediates between almost any two features a fungus can have. If it is not clear which branch to take on a page, you should try multiple options.

Take every identification with a grain of salt, even if the text uses the phrase "it is." Heed hedge phrases even more—for example, "might be" here means your find is likely to be one of *many* possibilities beyond those listed. **Don't eat anything** based on this guide alone.

Panus lecomtei

Suillus cavipes

Puccinia coronata

Stilbella fimetaria

Fungi come in many different shapes, some recognizably mushrooms, and others less so.

GLOSSARY

- A **fungus** is any organism placed in the kingdom *Fungi*. None of them can walk or photosynthesize, but they are really only united by descending from a certain branch of the evolutionary tree. They can be microscopic, form enormous mushrooms, or do almost anything in between.

It doesn't necessarily matter what a **mushroom** is. Everyone agrees that a mushroom *at least* has to be:

gall in oak leaf petioles, caused by
Andricus quercuspetiolicola wasps

- a structure formed by a fungus,
- not a mold or a typical lichen,
- and visible to the naked eye.

Opinions vary widely on how else, if at all, the organism should be restricted. Mushrooms and non-mushrooms are both included here. However, we will only indicate that a species is a mushroom if it also:

- is **at least 5 mm** tall, wide, or thick (as is everything covered by this book),
- has its own independent shape; i.e., is **not amorphous** (not a crust, lump, or bulge),
- is **not a gall** (a parasite-induced growth of an unrelated host's tissue),
- and is **not a sclerotium** (a dense, nonreproductive mass of fungal tissue).

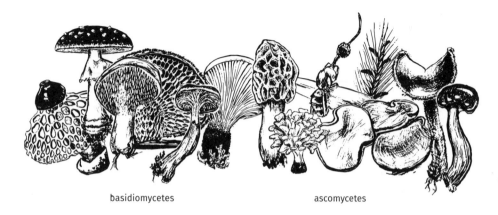

basidiomycetes   ascomycetes

- ***Basidiomycota*** and ***Ascomycota*** are the two largest phyla (major subdivisions) of kingdom *Fungi*, and nearly every mushroom you find will belong to one of them—it will be a **basidiomycete** or an **ascomycete**. The most consistent difference is in their microscopic spore-producing structures. You will not need to be able to tell the phyla apart to use this guide. However, you will eventually notice tendencies distinguishing them. For example, all mushrooms with gills, pores, or teeth are basidiomycetes.

- The **substrate** of a find is what it is growing on, in, or from. The most common substrates for mushrooms are on the ground (e.g., soil, leaf litter, wood chips, or grass) or on wood (e.g., small chunks, logs, or standing trunks).

mycelium within a substrate

- **Mycelium** is the root-like network of filaments or cords of a fungus, spreading and feeding inside the substrate, which does not produce spores itself, but eventually produces mushrooms. It is usually either microscopic or naked-eye visible and white. It can persist for years.

- An **effused** growth is flattish and overall spread, smeared, or splattered across the substrate.

Peniophora albobadia, an effused fungus

- A **spore print** is a visible accumulation of spores deposited by a mushroom. Its color is an important tool for identification within some categories. If you want to use this feature, you will have to obtain one yourself, unless you are lucky enough to find a natural spore print on something underneath the mushroom. Cut off the stem, place the cap on a piece of tin foil with one cap edge lifted up by a small object like a pencil, cover it all with a bowl, and leave it untouched for several hours.

With a certain amount of experience, you will be able to guess just by looking at most agarics whether they are dark-spored,

spore print method

5

pink-spored, or light-spored, and the spore print will become less crucial. This book will only ask for spore print color to reach some broad categories of agarics.

- In mycology, a **cap** is not just a structure perched on top of a stem. A shelf mushroom sticking out from wood, with no stem, has a cap. It *is* a cap!

Ganoderma applanatur a mushroom with a cap lacking a stem

- A surface of a mushroom is **zonate** if it has at least a few distinct, concentric bands of different hues.

- **Perithecia** are miniscule to microscopic flask-shaped divets, on some ascomycetes, that produce spores. On many species, they can be easily spotted with a hand lens or good vision as a regular pattern of well-separated, miniscule pimples or darkened dots.

a zonate cap

- A **lichen** is a living structure made of multiple species: a fungus farming algae and/or cyanobacteria. The fungus provides shelter and the other species provides nutrients by photosynthesis. Lichens are studied separately from mushrooms. They are:

perithecia, under a hand lens or microscope

  1. not bulky: either

  (a) ornately protruding from the substrate: leafily, densely branched, or beard-like; never with extensive smooth surfaces, or

    (b) spread along the substrate: ornately roughened and mosaic-like, or paint-like and dotted with little ornaments, more densely towards the middle.

2. tough or papery—not cottony, squishy, or fleshy like a mushroom.

3. quite dry (unless soaked by dew or rain, of course).

4. usually cool- or dull-colored, especially in gray to greenish shades, or bright yellows or oranges.

5. usually prominently exposed, especially on tree bark or fallen twigs, rocks, or wood in use in fences, etc.

You will quickly get a sense for what is a lichen and what is not.

Acarospora contigua,
a lichen

- A **slime mold** (or myxomycete, or myxo) is an amoeba visible to the naked eye. They look fungal but belong to another kingdom. Your find is probably a slime mold if any of these three apply:

1. It is a yellow or white "liquid" arranged in a fine network of thin veins.

Psora pseudorussellii,
a lichen

2. It is not a liquid but remarkably fragile, and a firm poke collapses it into a milky or densely powdery substance.

3. It is a troop of many small, distinctly evenly spaced individuals, each with one or more heads (<2 mm wide) perched on (or dangling from) a tiny (not wispy) stem.

Stemonitis, a slime mold

- **Achlorophyllous plants** are plants that don't photosynthesize and aren't green. Some can be confused for mushrooms, but they all have flowers, little shield-like leaves pressed against the stem, or both. If yours is entirely white to pinkish orange, shaped like a candy cane and/or has a flower at the tip, it is a **ghost pipe** (*Monotropa uniflora*) or **pinesap** (*M. hypopitys*). If it looks like a "corn cone cluster," each cone yellow and covered with dark orange-dipped leaves or flowers, it is a **cancer-root** (*Conopholis*).

Monotropa, an achlorophyllous plant

FINDER KEY

Most identifications suggested by this key are basidiomycete mushrooms. Those that are *not*—those that belong to other phyla, or that we don't consider to be mushrooms—are annotated with superscript letters. Superscripts are capitalized for mushrooms (ascomycetes[A], other fungi[F]) and lowercase otherwise (ascomycetes[a], basidiomycetes[b], other fungi[f], and non-fungi[x]). These will usually only appear at the end of the final choice. When a final choice has multiple suggestions with different statuses, each (non-basidiomycete or non-mushroom) name will be annotated separately.

Conopholis (cancer-root)

BEGIN HERE

Check the glossary to see if your find is an **achlorophyllous plant.**[x]

gills

Check the glossary to see if your find is likely to be a slime mold. If so,

GO TO PAGE 11

Look at the underside—the surface that was facing the ground (or as close as possible) *before* you found it. If the underside

pores

- has gills (many aligned "plates," "sheets," or "ridges," most often radiating out from a stem or base), it is an **agaric**.

GO TO PAGE 13

an agaric

tubelets

- is formed from a great many miniscule (≤1 mm), squished-together, open-mouthed cups or tubelets,

GO TO PAGE 18

- is formed from a great many small to tiny "holes" or "divets," they are pores. GO TO PAGE 19

teeth

If it has a cap whose edges are entirely connected to a stem, tear open a bit of the cap's underside to make sure there are no gills

(GO TO PAGE 13) or pores (GO TO PAGE 19) hidden by a veil.

Check the glossary to see if your find is likely to be a lichen. If so,

GO TO PAGE 24

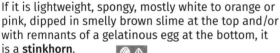
a lichen

If the underside is formed from a great many teeth ("spines," "prickles," or "shards" hanging downward),

GO TO PAGE 26

a toothed fungus

If it is lightweight, spongy, mostly white to orange or pink, dipped in smelly brown slime at the top and/or with remnants of a gelatinous egg at the bottom, it is a **stinkhorn**.

GO TO PAGE 27

a stinkhorn

If the interior and underside are rubbery-gelatinous (or, in dry conditions, like dried glue)

GO TO PAGE 29

a jelly

If it is growing from a bug (if the base is buried in soil or wood, dig it up carefully to check; the bug may be consumed by a soft oblong cottony pouch of mycelium), **GO TO** **PAGE 33**

Cordyceps tenuipes, on a buried cocoon

If it is growing on a fruit, flower, cone, leaf, or thin stem of a living plant, or swelling from inside a branch, **GO TO** **PAGE 33**

Apiosporina morbosa, growing on a cherry twig

If it looks like a little frilly bubblegum tuft (light pink, ~1 cm wide or less), on wood, with or without a short blackish stem, it is a **bubblegum xylaria** (anamorph stage of *Xylaria flabelliformis*).[A]

Otherwise, begin identifying simply by its overall shape. **GO TO** **PAGE 13**

Xylaria flabelliformis, bubblegum xylaria

 If it is a thin, complex network of veins of a milky liquid, it is the feeding stage (*plasmodium*) of a **slime mold**. It will be difficult to identify further.[x]

Otherwise, it probably is (or soon will be) the reproducing stage (*sporocarp*) of a **slime mold**. **GO TO SLIME A PAGE 12**

a slime mold, feeding stage

11

SLIME A

If it is uniformly white to yellow, and made of either short, upright, wavy tentacles, or porous, golf ball-like cushions with sharp ridges surrounding empty chambers, it is a **coral slime mold** (*Ceratiomyxa fruticulosa*).[x]

Hemitrichia serpula, pretzel slime mold

If it is a simple, bold network of snaking walls, it is a **network slime mold** forming a *plasmodiocarp*. It is probably made of orange tubes and in that case is a **pretzel slime mold** (*Hemitrichia serpula*).[x]

If it is cushion-shaped and made of a great number of tiny columns or nubs squished together, it is a **false-cushion slime mold** forming a *pseudoaethalium*—perhaps *Tubifera* or *Dictydiaethalium*.[x]

Tubifera, a false-cushion slime mold

If it is cushion-shaped otherwise, or up close it reminds you of a delicate heap of scrambled eggs, or it is a cleanly rounded blob or sphere, and it is

Inonotus rickii, rustburst

- rust-colored, stringy, and powdery, it is a **rustburst** (*Inonotus rickii*).[b]

- otherwise, it is a **cushion slime mold** forming an *aethalium*—perhaps *Fuligo*, *Didymium mucilago*, a **wolf's-milk slime mold** (*Lycogala*), or *Reticularia*.[x]

Lycogala, wolf's-milk slime mold

If it is a troop of many small individuals with one or more heads (<2 mm wide) perched on (or dangling from) a thin stem, and

Phleogena faginea, fenugreek stalkball

- it smells like maple syrup, it is the **fenugreek stalkball** (*Phleogena faginea*).

- otherwise, it is a **trooping slime mold** forming many *sporangia*—perhaps *Stemonitis*, *Stemonitopsis*, *Arcyria*, or *Badhamia*.[x]

Badhamia, a trooping slime mold

 If it has no stem, or only a stubby one at (or near) the edge of the cap, GO TO GILL A BELOW

If it has a distinct stem, attached somewhere near the middle of the cap, it is an **agaricoid**. GO TO GILL D PAGE 15

Schizophyllum commune, a splitgill

GILL A

If the edge of each gill (look closely) is split into two parallel edges, it is a **splitgill** (*Schizophyllum commune*).

Trametes betulina, gray polygill

If it has an entirely cardstock-like to woody consistency, and a zonate cap and/or pore-like or labyrinth-like sections in the gills, it is a **polygill**. It is probably a **gray polygill** (*Trametes betulina*),

13

rusty polygill (*Gloeophyllum sepiarium*), **blushing bracket** (*Daedaleopsis confragosa*), or **grand polygill** (*Fomitopsis quercina*).

Fomitopsis quercina, grand polygill

Otherwise, it is a **pleurotoid**. GO TO GILL B BELOW

GILL B

If the gills are contorted upwards, completely overtaking the cap, dingy whitish, puffy and crumpled, with the slots between partly filled with tissue, it is a **slipper** (*Crepidotus*) parasitized by *Hypomyces tremellicola*.[a]

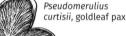

Phyllotopsis nidulans, mock-oyster

If it is quite evenly colored one shade between orangish yellow and salmon orange, and the cap is distinctly fuzzy (not just velvety), it is a **mock-oyster** (*Phyllotopsis nidulans*).

Pseudomerulius curtisii, goldleaf pax

If it is orange to olive-yellow and all of the gills are thick and strikingly wiggly, it is a **goldleaf pax** (*Pseudomerulius curtisii*).

Otherwise, obtain a spore print. GO TO GILL C BELOW

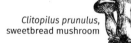

Clitopilus prunulus, sweetbread mushroom

GILL C

If the spore print is pink, it is a **pinkgill** in *Clitopilus* or *Entoloma* subgenus *Claudopus*.

If the spore print is brownish to blackish, it is a **dark-spored pleurotoid**. It is a **slipper** (*Crepidotus*), **eccentric olvet** (*Simocybe haustellaris* group), **flamester** (*Pleuroflammula flammea* group), or **fan pax** (*Tapinella panuoides*).

Pleurocybella porrigens, angel wing

If the spore print is pastel-colored to yellowish to whitish, it is a **light-spored pleurotoid**—perhaps an **oyster mushroom** (*Pleurotus*), **oysterling** (*Hohenbuehelia*), **angel wing** (*Pleurocybella porrigens*), **soft sawgill** (*Lentinellus*), or **gillchalice** (*Panus*).

Pleuroflammula flammea group, a flamester

GILL D

If it is not tiny, with shallow to vein-like, blunt-edged gills that are poorly distinguished from the cap and run down the stem, it is a **cantharelloid**. It is probably a **chanterelle** (*Cantharellus*), **trumpet** or **yellowfoot** (*Craterellus*), **gomphelle** (*Gomphus* or *Turbinellus*), *Polyozellus*, or **Aphrodite's veinlet** (*Aphroditeola olida*).

Cantharellus, a chanterelle

Polyozellus

If it is growing on wood and has a strikingly netted-wrinkled pink cap, pink gills, and a stout whitish stem, it is a **wrinkled-peach mushroom** (*Rhodotus*).

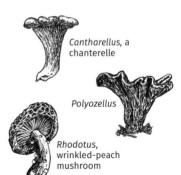

Rhodotus, wrinkled-peach mushroom

If the cap is ≤2 cm with a distinctly powdery gray-to-brown color with fluff hanging from the edge, and gills of reddish pink to pinkish brown to bloody brown that do not touch the stem, it is a **lamprey powderling** (*Melanophyllum*).

If the cap is ≤3 cm, with a matte, solid-colored mahogany to blackish-brown color, distinctly fading to a light yellowish glow at the edges, and smells of cucumber or fish, it is a **cucumbercap** (*Macrocystidia*).

Otherwise, obtain a spore print. GO TO GILL E BELOW

GILL E

If the spore print is yellow to brown, and

- the gills are concentric, it is a **labyrinth tigereye** (*Coltricia montagnei* var. *greenei*).

- the stem is thick, dark brown, and fuzzy, it is a **velvet-footed pax** (*Tapinella atrotomentosa*).

- the gills are thick, bright yellow, and decurrent, it is a *Phylloporus*.

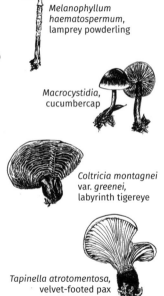

Melanophyllum haematospermum, lamprey powderling

Macrocystidia, cucumbercap

Coltricia montagnei var. *greenei*, labyrinth tigereye

Tapinella atrotomentosa, velvet-footed pax

- the cap is >5 cm, there is a large veil, and cap and stem are evenly caramel-colored with a fine granular coat, it is a **golden bootleg** (*Phaeolepiota aurea*).

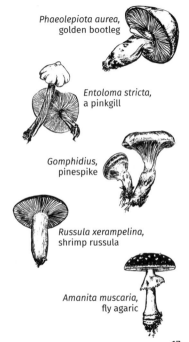

Phaeolepiota aurea, golden bootleg

If the spore print is green, it is a **green parasol** (*Chlorophyllum molybdites*).

If the spore print is distinctly (or brownish-) pink, it is either a **shield**, **sheath**, or **sword** (*Pluteaceae*); **pinkgill** (*Entolomataceae*); or a *Lepista*.

Entoloma stricta, a pinkgill

If the spore print is brownish to blackish, it is a dark-spored agaricoid—perhaps a **fieldcap** (*Agrocybe*), **inkcap**, **webcap** (*Cortinarius*), **pinespike** (*Gomphidius*), **tuft** (*Hypholoma*), or **pan** (*Panaeolus*), but it is probably one of countless others.

Gomphidius, pinespike

If the spore print is pastel-colored to yellowish to whitish, it is a light-spored agaricoid—perhaps a **milkcap** (*Lactarius* or *Lactifluus*), **russula** (*Russula*), **amanita** (*Amanita*), **waxcap**, **parasol** (*Chlorophyllum*, *Macrolepiota*, or *Leucocoprinus*)**,** or **rooter**, but it is probably one of countless others.

Russula xerampelina, shrimp russula

Amanita muscaria, fly agaric

If it is effused, GO TO TUBELET A BELOW.

If it has a cap on a quite distinct stem, it is *Pseudofistulina radicata*.

If it has a cap but no stem, or only a stubby one at the edge of the cap, it is a **beefsteak polypore** (*Fistulina americana*).

Pseudofistula radicata

Fistulina americana, beefsteak polypore

TUBELET A

If the cups are bright orange to yellow, and on living leaves, fruits, or twigs, they are shallow **rust pustules** (*aecia*) formed by a **rust fungus**. Depending on the host, it might be **mayapple rust** (*Allodus podophylli*), **blackberry rust** (*Gymnoconia peckiana*), **jack-in-the-pulpit rust** (*Uromyces ari-triphylli*), or a *Puccinia*.[b]

If the cups are all the same width, and are thin tubes or (almost) entirely adjoined, it is either a **tubelet** (possibly in *Henningsomyces*, *Rectipilus*, or *Merismodes*) or a **white cupcrust** (*Porotheleum fimbriatum*).[b]

Allodus podophylli, mayapple rust

Gymnoconia peckiana, blackberry rust

Porotheleum fimbriatum, white cupcrust

If it is a stout, lumpy, or pointy mass; is entirely covered with pores, and;

- it has pink to red tones, it is a **bleeding rosette** (*Abortiporus biennis*).[b]

- it has orange tones, it is an **orange-sponge polypore** (*Pycnoporellus alboluteus*).[b]

If it is effused, it is a **porecrust**—perhaps an *Antrodia*, *Ceriporia*, *Fomitiporia*, *Pycnoporellus*, or *Xylodon*, but it is probably one of countless others.[b]

If it has one or more obvious stems, attached somewhere near the middle of the cap, GO TO PORE A BELOW

If it has no stem, or only a stubby one at (or near) the edge of the cap, it is a **bracket polypore**. GO TO PORE C PAGE 23

Abortiporus biennis, bleeding rosette

Pycnoporellus alboluteus, orange-sponge polypore

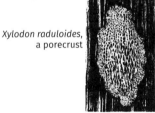

Xylodon raduloides, a porecrust

PORE A

If it has a typical, fleshy mushroomy consistency, and the pore layer can be easily separated from the cap, it is a **bolete**. GO TO BOLETE A PAGE 20

19

If it has a woody to leathery, hard-to-tear consistency, and the pore layer cannot be easily separated from the cap, it is a **stalked polypore**. GO TO PORE B PAGE 22

BOLETE A

If the stem is

- at least partly covered with distinct, tiny dots, it is probably a **jack** (*Suillus*), **scaberstalk** (*Leccinum*), *Hemileccinum*, **chrome-footed bolete** (*Harrya chromipes*), or *Sutorius eximius*.

- at least partly covered with a network of strikingly protruding ridge-like strips (not just veins), it is a **shagstalk** in *Aureoboletus*, **Frost's bolete** (*Exsudoporus frostii*), or **moon bolete** (*Austroboletus subflavidus*).

- hollow, it is a *Gyroporus* or a **hollow jack** (*Suillus ampliporus*).

Otherwise, GO TO BOLETE B BELOW

BOLETE B

If the stem has a ring or distinct ring-like zone, there are veil remnants hanging from the cap edge, or the cap has striking

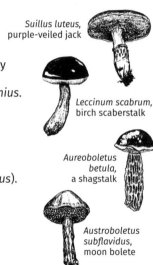

Suillus luteus,
purple-veiled jack

Leccinum scabrum,
birch scaberstalk

Aureoboletus betula,
a shagstalk

Austroboletus subflavidus,
moon bolete

scales or feels slimy or tacky, it is probably a **jack** (*Suillus*), **powdery-sulphur bolete** (*Pulveroboletus ravenellii*), *Strobilomyces,* **old-man-of-the-woods** (*Strobilomyces*), or **pineapple bolete** (*Boletellus ananas*).

Strobilomyces, old-man-of-the-woods

If it is growing on less-than-thoroughly-rotted wood, wood chips, or an earthball, it is probably a *Buchwaldoboletus*, **svelte bolete** (*Boletellus chrysenteroides*), **admirable bolete** (*Aureoboletus mirabilis*), **bay bolete** (*Imleria badia*), or **parasitic bolete** (*Pseudoboletus parasiticus*).

Boletellus ananas, pineapple bolete

Buchwaldoboletus

Otherwise, GO TO BOLETE C BELOW

BOLETE C

If the pores are

- orange to red, it might be a *Neoboletus*, *Rubroboletus*, *Suillellus*, or *Lanmaoa*.

Rubroboletus eastwoodiae, Satan's bolete

- distinctly radially arranged, or distinctly yellow, angular, and large (<40 from stem to mature cap edge), it might be a *Xerocomus*, *Xerocomellus*, **ash-tree bolete** (*Boletinellus merulioides*), **goldbolete** (*Aureoboletus*), or *Hortiboletus*.

Boletinellus merulioides group, ash-tree bolete

- distinctly yellow, and stain blue when rubbed, it might be a **bicolor bolete** (*Baorangia*), *Lanmaoa*, **butter bolete** (*Butyriboletus*), or **bitter bolete** (*Caloboletus*).

Otherwise, it might be a *Boletus*, *Tylopilus*, *Retiboletus*, or *Chalciporus*.

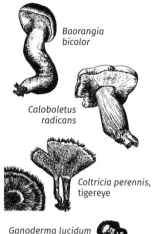

Baorangia bicolor

Caloboletus radicans

Coltricia perennis, tigereye

Ganoderma lucidum group, reishi

PORE B

If it is growing on the ground, is mostly brown to brownish orange, and has a distinctly round and distinctly zonate cap, it is a **tigereye** (*Coltricia*).

If it is a bouquet of more than a dozen caps, it is an **umbrella polypore** (*Polyporus umbellatus*).

If both the cap and stem are hard and ~smooth (possibly rippled), it is a stalked conk. It is a **reishi** (*Ganoderma*), *Amauroderma*, or **spinning-plate conk** (*Microporellus dealbatus*).

Otherwise, it is a typical stalked polypore—perhaps a *Cerioporus*, *Picipes*, *Lentinus*, **kurotake** (*Boletopsis*), or an **albatrelle** (*Scutigeraceae*).

Boletopsis, kurotake

PORE C

If it is on the ground or the base of a tree, and round or forming many-capped rosettes, it is probably a **Berkeley's polypore** (*Bondarzewia berkeleyi*), **blackening polypore** (*Meripilus sumstinei*), **dyer's polypore** (*Phaeolus schweinitzii* group), *Laetiporus cincinnatus*, *Kusaghiporia persicina*, **velvet rosette** (*Onnia*), **hen-of-the-woods** (*Grifola frondosa*), or **bleeding rosette** (*Abortiporus biennis*, SEE PAGE 19).

Otherwise, it is a **shelf polypore**. GO TO PORE D BELOW

Bondarzewia berkeleyi, Berkeley's polypore

Grifola frondosa, hen-of-the-woods

PORE D

If it is >3 cm wide; robust or swollen; cleanly formed (not smeared down the tree), with a ~hard, smooth, or woody (perhaps rippled or cracked) cap surface; and makes a satisfying noise (perhaps a "bonk") when you rap it with your knuckles, it is a **conk**—perhaps a **reishi** or **artist's conk** (*Ganoderma*), **beltconk** (*Fomitopsis*), *Fomes*, *Fulvifomes*, or *Phellinus*.

If it is thick and distinctly squishy, it is probably a **cheeseshelf** (perhaps *Postia* or *Tyromyces*), **bear bracket** (*Amylocystis lapponica*), **sherbet bracket** (*Leptoporus mollis*), or **climax bracket** (*Climacocystis borealis*).

Phellinus

Climacocystis borealis, climax bracket

23

If it is entirely, top and bottom, a few shades of brown (light-, gray-, golden-, rusty-, dark-), perhaps partly dark reddish brown or blackish on top, and quite firm, it is probably a polypore in *Hymenochaetaceae*—perhaps a **mustard bracket** (*Fuscoporia gilva*), **shaggy bracket** (*Inonotus hispidus*), **radiated bracket** (*Mensularia radiata*), or a **pawpaw bracket** (*Phylloporia amplectens*).

Otherwise, it is a relatively typical bracket—perhaps a *Trametes*, a **violettooth** (*Trichaptum*), a **honeycomb polypore** (*Neofavolus alveolaris* group), a **mossy-maze bracket** (*Cerrena unicolor*), a **smoky bracket** (*Bjerkandera*), or a **chicken-of-the-woods** (*Laetiporus*), but it is probably one of countless others.

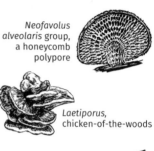

Neofavolus alveolaris group, a honeycomb polypore

Laetiporus, chicken-of-the-woods

Ignore "ornaments" that are a different color and texture than the "base layer."

If it is in sheets of loose netting, it is a **lace lichen** (*Ramalina menziesii*).[a]

If it is beard-like, made of "hairs," it is probably a **beard lichen** (*Usnea*), **witch's hair** (*Alectoria sarmentosa*), or **horsehair lichen** (*Bryoria*).[a]

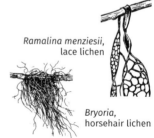

Ramalina menziesii, lace lichen

Bryoria, horsehair lichen

If it is many-branched or has obvious columns or tentacles, it is a **fruticose lichen**—perhaps a *Cladonia*, **oakmoss** (*Evernia*), *Ramalina*, or **wolf lichen** (*Letharia*).[a]

Letharia columbiana, wolf lichen

If it is formed of "leaves" easily liftable from the substrate, it is a **foliose lichen**—perhaps a **greenshield lichen** (*Flavoparmelia*), *Lobaria*, **rocktripe** (*Umbilicaria*), or **pelt lichen** (*Peltigera*).[a]

Xanthomendoza, a sunburst lichen

If it is formed from tiny leaf-like plates firmly hugging the substrate, it is a **squamulose lichen**—perhaps a **sunburst lichen** (*Xanthomendoza*), **elf-ear lichen** (*Normandina pulchella*), **tree stipplescale** (*Placidium arboreum*), or a **fishscale lichen** (*Psora*).[a]

Placidium ardoreum, tree stipplescale

If it is simply powdery, it is a **leprose lichen**. It is probably a **dust lichen** (*Lepraria*) or **gold-dust lichen** (*Chrysothrix*).[a]

Lepraria, a dust lichen

Otherwise, if it is more or less flat or mosaic-like, it is a **crustose lichen**—perhaps a **Christmas lichen** (*Herpothallon rubrocinctum*), **speckled blister lichen** (*Viridothelium virens*), **boulder lichen** (*Porpidia*), or **cobblestone lichen** (*Acarospora*).[a]

Herpothallon rubrocinctum, Christmas lichen

Viridothelium virens, speckled blister lichen

If it is gelatinous, it is a **jellyhog** (*Pseudohydnum*).

If it has a cap or branches, GO TO TOOTH A BELOW

If it is effused, GO TO TOOTH B PAGE 27

Pseudohydnum, jellyhog

TOOTH A

If it is not growing from wood and has a distinct stem, it might be a **hedgehog mushroom** (*Hydnum*), **earpick mushroom** (*Auriscalpium vulgare*), *Hydnellum*, or *Sarcodon*.

If it is growing from wood, entirely whitish, has large teeth, and there is no abrupt distinction between the teeth and the top of the "cap," it is a *Hericium*. It is probably a **lion's mane** (*H. erinaceus*), **bear's head** (Western *H. abietis*, Eastern *H. americanum*), or *H. coralloides*.

If it has no stem, or only a stubby one at (or near) the edge of the cap, and is growing from wood, it might be a **Northern tooth** (*Climacodon septentrionalis*), **Southern tooth** (*Donkia pulcherrima*), an aging **violettooth** (*Trichaptum*), or a **marshmallow polytooth** (*Irpiciporus mollis*).

Hydnum, hedgehog mushroom

Pseudohydnum, jellyhog

Hericium erinaceus, lion's mane

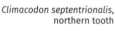

Climacodon septentrionalis, northern tooth

Irpiciporus mollis, marshmallow polytooth

TOOTH B

If the teeth all emerge from a shared layer spread along wood, it is a **toothcrust**—perhaps an **Asian beauty** (*Radulomyces copelandii*), **olive toothcrust** (*Hydnoporia olivacea*), **flaming toothcrust** (*Hydnophlebia chrysorhiza*), or *Steccherinum*.[b]

Radulomyces copelandii, Asian beauty

Hydnophlebia chrysorhiza, flaming toothcrust

If the teeth are separate and bright orange, on living leaves, fruits, or twigs, they are not actually teeth. They are tube-shaped **rust pustules** (*aecia*) packed with spores—perhaps **cedar-quince rust** (*Gymnosporangium clavipes*), **crown rust** (*Puccinia coronata*) on oat or barley, or any of many other *Puccinia* species.[b]

Gymnosporangium clavipes, cedar-quince rust

If the teeth are separate and growing on wood, it is a **spineswarm** (*Mucronella*).[b]

Mucronella pendula

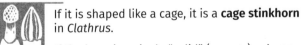

If it is shaped like a cage, it is a **cage stinkhorn** in *Clathrus*.

If the base is a single "solid" (spongy) column, short or tall, GO TO STINKHORN A PAGE 28

Clathrus ruber, a cage stinkhorn

If the base is two to several "solid" (spongy) columns, GO TO STINKHORN B PAGE 28

STINKHORN A

Aseroe rubra,
anemone stinkhorn

If it has several reddish, pointed "tentacles," each split like a snake's tongue, sticking straight out from a short column base, it is an **anemone stinkhorn** (*Aseroe rubra*).

If it has a narrow, cone-like, pointed muddy tip, it is either a **dog stinkhorn** (*Mutinus*) or a **wolf stinkhorn** (*Phallus rubicundus* group).

Mutinus, dog
stinkhorn

If it has a stout muddy tip, riddled with shallow pits and/or with a flattened top, or has an large, elegant, doily-like veil, it is a **phallus stinkhorn** (*Phallus*).

Otherwise, it is a **lizard stinkhorn** (*Lysurus*).

STINKHORN B

If the columns are splayed outwards, it is a **devil's-fingers stinkhorn** (*Clathrus archeri*).

Clathrus archeri,
devil's-fingers stinkhorn

If the columns are fused together at the tips, it is probably a **column stinkhorn** (*Clathrus columnatus*), **squid stinkhorn** (*Pseudocolus fusiformis*), **jewel stinkhorn** (*Laternea dringii*), or **ivory stinkhorn** (*Blumenavia rhacodes*).

Clathrus columnatus,
column stinkhorn

Laternea dringii,
jewel stinkhorn

If it has an indistinct shape—a convoluted, wrinkly, or brain-like mass, *without* smooth (flat or concave) surfaces, **GO TO JELLY A BELOW**

If it has a distinct shape, **GO TO JELLY D PAGE 31**

JELLY A

If it is a color of the rainbow, **GO TO JELLY B BELOW**

Otherwise, **GO TO JELLY C PAGE 30**

JELLY B

If it is red, it is **red witches'-butter** (*Tulasnella aurantiaca*).[b]

If it is orange to yellow, it is a basidiomycete. It is probably **yellow witches'-butter** (*Tremella mesenterica*), **golden witches'-butter** (*Naematelia aurantia*), **conifer witches'-butter** (*Dacrymyces chrysospermus* group), or a **rust** in *Gymnosporangium*.[b]

If it is green and on

* the ground, it is a **common nostoc** (*Nostoc commune*), a bacterial colony.

Tulasnella aurantiaca, red witches'-butter

Naemantelia aurantia, golden witches'-butter

Nostoc commune, common nostoc

29

• wood, bark, or a rock, it is a **jelly lichen** (*Collema* or *Leptogium*).[a]

If it is pinkish to purplish, it is a **jellydisc** (*Ascocoryne* or *Neobulgaria*).[a] GO TO JELLY G PAGE 32

JELLY C

If it is growing from the top of a common coincap (*Gymnopus dryophilus*), it is a **coincap cloud** (a gall caused by *Syzygospora mycetophila*).[b]

Syzygospora mycetophila,
a gall on *Gymnopus*
dryophilus, common coincap

If it is richly amber-colored, it is a **jellyroll** (*Exidia recisa* group) or **jellyleaf** (*Phaeotremella*).[b]

Phaeotremella,
jellyleaf

If it is pale amber-colored to buff to gray, it is a **crystal brainjelly** (*Myxarium nucleatum* group) or **nougat brainjelly** (*Naematelia encephala*).[b]

If it is whitish, it is **white witches'-butter** (*Ductifera pululahuana*), **snow jelly** (*Tremella fuciformis*), or **tube-sock jelly** (*Helvellosebacina concrescens* group).[b]

Ductifera pululahuana,
white witches'-butter

If it is black, it is **black witches'-butter** (*Exidia glandulosa* group).[b]

Tremella fuciformis,
snow jelly

JELLY D

If it is ~amber~ to dull flesh-colored and/or cup/disc-shaped, GO TO JELLY E BELOW

If it is entirely yellow to orange to pink, GO TO JELLY F PAGE 32

Otherwise, GO TO JELLY G PAGE 32

JELLY E

If it is amber and

- has a stem, it is a **velvet earlet** (*Dacryopinax elegans*).

- large (≥1.5 cm across), thin (like a human ear or thinner), and the top and bottom have a different texture and luster, it is a **jelly ear** (*Auricularia*).

- small (≤2.5 cm across), thick (like a peppermint patty or thicker), and the top and bottom have a nearly identical texture and luster, it is a **jellyroll** (*Exidia recisa* group).

Otherwise, it is a jelly cup—perhaps a **corticioid jellyspot** (*Dacrymyces corticioides*)[b], **ebony cup** (*Pseudoplectania nigrella*)[A], **black bulgar** (*Bulgaria inquinans*)[A], or **jellydisc** (*Ascocoryne* or *Neobulgaria*)[A].

Dacryopinax elegans, velvet earlet

Auricularia, jelly ear

Pseudoplectania nigrella, ebony cup

Bulgaria inquinans, black bulgar

31

JELLY F

If there are several congregated on dead wood, and they are

- greenish yellow, it is a **gluedrip** (*Gloeomucro*).

- otherwise, it is probably a *Dacrymyces*, **jellycone** (*Guepiniopsis*), or *Calocera*.

Otherwise, it is probably a *Guepinia*, *Gymnosporangium*[b], or a different *Calocera*.

Gloeomucro dependens, pendant gluedrip

JELLY G

If it is growing on wood and is a cluster of nodules or swollen discs, it is a **jellydisc** (*Ascocoryne* or *Neobulgaria*).[A]

If it has a distinct cap and stem, it is a **jellybaby** (*Leotia*).[A]

If it is a complex ball of many nubs or tubes pointing outwards, it is a **jellyflower** (*Sebacina sparassoidea*).

If it is club-shaped and growing in moss, it is a **moss jellyclub** (*Eocronartium muscicola*).

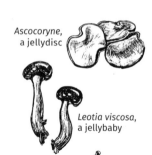

Ascocoryne, a jellydisc

Leotia viscosa, a jellybaby

Sebacina sparassoidea, jellyflower

If it is a pale, crumbly, dusty "plug" filling the end of a cicada abdomen, it is a **cicada destroyer** (*Massospora*).[f]

If it is one to several dozen spikes, tentacles, clubs, or branched structures, it is a **cordycipitoid** in *Akanthomyces*, *Cordyceps* (page 11), *Gibellula*, *Ophiocordyceps*, *Paraisaria*, *Pleurocordyceps*, or *Purpureocillium*.[A]

If it is mold-like (dusty or fuzzy material overtaking the bug, perhaps swelling it), it might be a *Beauveria*, *Erynia*, *Engyodontium*, *Metarhizium*, or *Purpureocillium*.[a]

 If it is on stork's-bill or crane's-bill leaves and made of a great many distinct, miniscule, blood-red, shiny spherical beads, they are galls formed by a **stork's-bill chytrid** (*Synchytrium papillatum*) or **crane's-bill chytrid** (*Synchytrium geranii*).[b]

If it is on sphagnum moss and is a yellow, well-formed blob, it is a **fractal moldtruffle** (*Endogone pisiformis*).[F]

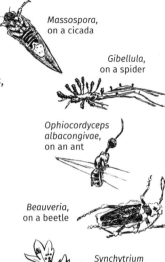

Massospora, on a cicada

Gibellula, on a spider

Ophiocordyceps albacongivae, on an ant

Beauveria, on a beetle

Synchytrium papillatum, a gall

If it is on or under a beech, spongy, and black or dingy yellowish tan, it is an **aphid sootsponge** (*Scorias spongiosa*).[a]

Scorias spongiosa, aphid sootsponge

If it is on a thin stem of *Prunus* (cherry, plum, etc.) and is hard and black, it is a **black knot** (*Apiosporina morbosa*).[a]

If it is a waxy, messy, continuous, white-to-pale-tan crust, it is an **encrusting waxspill** (*Sebacina incrustans*).[b]

Apiosporina morbosa, black knot

If it is dangling (or fallen) from a redcedar, nearly spherical, mahogany- to burgundy-colored, hard, smoothish, and evenly pitted or lumpy all the way around, it is a gall caused by **cedar-apple rust** (*Gymnosporangium juniperi-virginianae*).[b]

If it is on stems of huckleberry or blueberry (*Vaccinium*), enveloping them in rusty tubes, it is **huckleberry broom rust** (*Calyptospora*).[b]

Gymnosporangium juniperi-virginianae, cedar-apple rust

If it is powdery, and

- dark chocolate brown, and on poison ivy or yellow trumpetbush, it is **poison ivy rust** (*Pileolaria brevipes*) or **yellow trumpetbush rust** (*Prospodium transformans*).[b]

- yellow to orange, it is a different **rust**—perhaps *Gymnoconia*, *Cronartium*, *Phragmidium*, or *Puccinia*.[b]

Pileolaria brevipes, on poison ivy

If it is on *Poales* (e.g., grass, wheat, bamboo), GO TO GALL A PAGE 36

If it is rough and woody; distinctly fuzzy, furry, spiky, or spike-like; or quite smooth and spherical, it is not fungal. It is probably a gall, caused by a **bacterium** (e.g., *Agrobacterium*, *Allorhizobium*), a **mite** (e.g., *Aceria*), or an **insect**: a wasp (e.g., *Amphibolips*, *Callirhytis*), a midge (e.g., *Rhopalomyia*, *Neolasioptera*), an aphid (e.g., *Mordwilkoja*, *Tetraneura*), a phylloxerid (e.g., *Phylloxera*), or a psylloid (e.g., *Pachypsylla*). It could also be a **gall-like scale insect** itself (e.g., *Allokermes*, *Nanokermes*).[x]

Otherwise, it might be fungal. It is probably a gall, caused by a **fungus** (*Taphrina*[a], *Exobasidium*[b], or *Melanopsichium*[b]), a **plasmodiophore** (*Sorosphaerula*)[x], a **mite** (e.g., *Aculops*, *Aceria*)[x], or an **insect**: a wasp (e.g., *Disholcaspis*, *Diastrophus*)[x], a sawfly (*Euura*)[x], a midge (e.g., *Polystepha*, *Rhopalomyia*)[x], an aphid (e.g., *Melaphis*, *Colopha*)[x], a phylloxerid (e.g., *Phylloxera*)[x], or a psylloid (e.g., *Pachypsylla*, *Trioza*)[x]. It could also be a **gall-like scale insect** itself (e.g., *Allokermes*, *Nanokermes*)[x].

Agrobacterium radiobacter, crown gall

Vasates quadripedes, an Eriophyidae mite gall

Besbicus mirabilis, a Cynipidae wasp gall

Taphrina alni, a fungal gall on alder cones

Eriophys laevis, an Eriophyidae mite gall

Mikiola fagi, a Cecidomyiidae midge gall

GALL A

If it is pink, it is **red thread disease** (*Laetisaria fuciformis*) or **pink patch disease** (*L. roseipellis*).[b]

If it is yellowish orange, it is a **clavicipitoid** in *Epichloe*.[a]

If it is on corn, shaped like a deformed garlic clove, whitish gray on the outside, and black inside, it is a gall caused by **corn smut** (*Mycosarcoma maydis*).[b]

Otherwise, it is another **smut** (Ustilaginales; e.g., *Ustilago, Testicularia*)[b] or a **clavicipitoid** in Clavicipitaceae (e.g., *Claviceps, Echinodothis*)[a].

Mycosarcoma maydis, corn smut

Claviceps purpurea, rye ergot

If it is simply star-shaped, with no extra structure in the middle, and the "rays" are

- short and unsightly, it is a post-mature **star earthball** (*Scleroderma polyrhizum*).

- slender and peachy, it is a **Texas star** (*Chorioactis geaster*).[A]

Ustilago nuda, barley smut

If it is shaped like **a disc or a cup** (even if quite shallow, upside-down, atop a stem, or many squished together), GO TO ◯ PAGE 37

Chorioactis geaster, Texas star

If it has a **distinct cap** (either atop a stem or simply as a horizontal shelf),

GO TO PAGE 39

If it is upright and **narrow**, club-shaped, or drop-shaped, at most
with an *indistinct* "cap" (a widened extension of the stem), or
many-branched with narrow or pointed tips, GO TO PAGE 44

If it is **effused**, GO TO PAGE 46

Hypoxylon crocopeplum, an effused fungus

If it is any stout, **well-formed** shape, independent of the substrate—a
sphere, blob, lozenge, perhaps distorted, but not flattened down,

GO TO PAGE 51

Sparassis crispa, a well-formed fungus

If it is **in between**—like a bulge, a flattened cushion, or a
head of hair—with substantial volume to it but still clearly
somewhat conforming to the substrate, GO TO PAGE 54

Aleuria aurantia, a cup fungus

If it is shaped like a trumpet or a martini glass or the opening
of one, and it is tough like paper or leather, it is not actually a cup
fungus (spores produced on inside surface). It is a *stereoid* fungus
(spores produced on outside surface).

GO TO PAGE 39

If it has a stem, it is a **stalked cup**—perhaps a **devil's urn** (*Urnula craterium* group)[A], **stalked scarlet elfcup** (*Sarcoscypha occidentalis*)[A], **felt saddle** (*Helvella macropus* group)[A], **sucker cup** (*Tatraea macrospora*)[A], but it is probably one of countless others.

Otherwise, GO TO CUP A BELOW

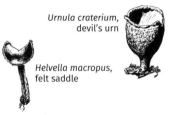

Urnula craterium,
devil's urn

Helvella macropus,
felt saddle

CUP A

If it has one or several tiny "eggs" or "lentils" inside, it is a **bird's-nest fungus** (*Nidulariaceae*).

Cyathus striatus,
fluted bird's-nest

If it is small (<1 cm across) and tough like cardstock, GO TO CUP B PAGE 39

If it is large (>5 cm across), powdery, and growing in grass or a similar open area, it is the remains of a certain kind of **puffball** (*Calvatia cyathiformis* group).

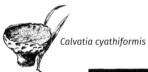

Calvatia cyathiformis

Otherwise, it is a **cup/disc fungus**—perhaps a **lemon disco** (*Calycina citrina*)[A], **woodstain elfcup** (*Chlorociboria*)[A], **orangepeel cup** (*Aleuria aurantia*)[A], or a **graceful lashcup** (*Humaria*)[A], but it is probably one of countless others.

Chlorociboria,
a woodstain elfcup

CUP B

If it is zonate inside or shallow and bleached, it is the cup-like appendage of a polypore, a **bird bracket** (*Trametes conchifer*), which may not be present.

Trametes conchifer, bird bracket

Otherwise, it is an emptied **bird's-nest mushroom** (*Nidulariaceae*).

Nidula, a bird's-nest mushroom

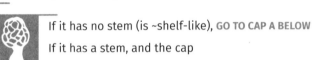

If it has no stem (is ~shelf-like), GO TO CAP A BELOW

If it has a stem, and the cap

- has a clearly distinct top side and underside, GO TO CAP B PAGE 40
- does not, GO TO CAP C PAGE 41

Stereum

CAP A

If it is thin, tough, leathery, or even hard, it is probably a *Stereum*, *Xylobolus*, **hoof stairstep** (*Hymenochaete rubiginosa*), **willow-glue stairstep** (*Hydnoporia tabacina*), or **bacon stairstep** (*Punctularia strigosozonata*).

Hymenochaete rubiginosa, hoof stairstep

If it is thick, fleshy, or soft, it is probably a **phlebie** (*Phlebia* or *Byssomerulius*), **silverleaf fungus** (*Chondrostereum purpureum*), **earthfan** (*Thelephora*), or **shallow mossmaid** (*Arrhenia retiruga*).

Phlebia tremellosa,
trembling phlebie

CAP B

If the stem does *not* widen trumpet-like into the cap, and the cap is a somewhat brittle, crumpled or folded sheet, it is a **saddle mushroom** (*Helvella*) or **lorchel** (*Gyromitra*).[A]
GO TO CAP F PAGE 43

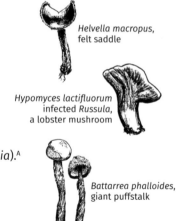

Helvella macropus,
felt saddle

If the underside of the cap has perithecia, it is a **lobster mushroom** (if bright orange) or a **gillgobble** (otherwise), in *Hypomyces*[a], parasitizing a **russula** or **milkcap**.

Hypomyces lactifluorum
infected *Russula,*
a lobster mushroom

If the top of the cap is distinctly convex, and the stem is

- strikingly dry and blackish, it is a **nail mushroom** (*Poronia*).[A]

- strikingly dry and woody-fibrous, it is a **giant puffstalk** (*Battarrea phalloides*).

Battarrea phalloides,
giant puffstalk

- typical fleshy mushroom-like, it is a **powdery starbear** (*Asterophora lycoperdoides*).

If it is fleshy like an ordinary mushroom, not tough, and

- all-white and small (stem <5 mm thick), it is **Humphrey's wing** (*Stereopsis humphreyi*).

- otherwise, it is either a **smooth chanterelle** (*Cantharellus lateritius* group), or a **trumpet** or **yellowfoot** (*Craterellus*).

Cantharellus lateritius group, smooth chanterelle

Otherwise, it is a stipitate stereoid. It is probably a **silver demurette** (*Stereopsis hiscens*), **diaphanous demurette** (*Cotylidia diaphana*), **wine-glass demurette** (*Podoscypha*), **lilac ribchalice** (*Cymatoderma caperatum*), or **earthfan** (*Thelephora*).

Craterellus tubaeformis, a yellowfoot

Cymatoderma caperatum

CAP C

If it is hard, entirely black to burgundy to brown, and growing on wood, it is a **King Alfred's cake** (*Daldinia*), probably *D. vernicosa* or *D. asphalatum*.[A]

Stereopsis hiscens, silver demurette

If there are many clustered on wood, entirely whitish, and the caps look like deflated bladders, they are **bladderstalks** (*Physalacria inflata*).

Physalacria inflata, bladderstalks

If the cap has perithecia, it is either a **truffleclub** (*Tolypocladium*)[A], **moonclub** (*Trichoderma leucopus* or *T. alutaceum*)[A], or a **cordycipitoid** (SEE PAGE 33).

If it is growing from soil; largely dingy whitish; with a distinctly phallic shape and sloppy, moldy texture; and with a roughly thimble-shaped cap on a stem girdled with sloppy cracks, it is an **amanita** parasitized by *Hypomyces hyalinus*[a].

Amanita parasitized by *Hypomyces hyalinus*

If the cap is neatly rounded or only slightly deformed, and has a "skin" filled with

- a spongy mass of heavily contorted little plates or chambers, it is a **secotioid**—perhaps a **tiger sawgill** (*Lentinus tigrinus*), **shaggy globe** (*Agaricus deserticola*), secotioid **russula** (several *Russula* species), or **wintergreen slickpouch** (*Pholiota nubigena*).

Pholiota nubigena, wintergreen slickpouch

- a distinctly marshmallow-like or powdery material, it is a **stalked puffball**. It is either a member of *Lycoperdaceae*, **shaggy pod** (*Podaxis*), **hotlips** (*Calostoma*), or a **fenugreek stalkball** (*Phleogena faginea*).

Lycoperdon (in *Lycoperdaceae*)

Phleogena faginea, fenugreek stalkball

Otherwise, it is an ascomycete. GO TO CAP D PAGE 43

CAP D

If it has a simple, rounded, smoothish cap, perhaps a bit squished or deformed, GO TO CAP E BELOW

If it has a more complex cap, GO TO CAP F BELOW

CAP E

If it is growing in water or wet mud, it is a **bogbeacon** (*Mitrula elegans* group) or **water club** (*Vibrissea truncorum*).[A]

If the cap is <5 mm wide, it is an **oakpin** (*Cudoniella acicularis*), **springpin** (*C. clavus*), **spruce beacon** (*Heyderia abietis*), or **Holway's beacon** (*Holwaya*).[A]

If the cap is ≥5 mm wide, it is a **rubberbaby** (*Cudonia*).[A]

CAP F

If it has quite distinct (not wrinkle-like) ridges and pits, it is a **morel** (*Morchella*).[A]

If it is distinctly thimble-shaped, smooth or wrinkly, it is a **thimblecap** (*Verpa*).[A]

Mitrula elegans group, bogbeacon

Vibrissea truncorum, water club

Holwaya mucida, Holway's beacon

Cudonia spathulata, manzanita rubberbaby

Morchella americana, a yellow morel

43

If it is saddle-shaped to heavily contorted or brain-like, it is a **saddle mushroom** (*Helvella*) or **lorchel** (*Gyromitra*).[A]

Gyromitra, a lorchel

 If it is dry; wiry to tough to hard; entirely black, gray, white, and/or beige; and it is

- dusty *all* over, it is the anamorph stage of *Xylaria poitei* or *X. allantoidea*.

- not, it is a **xylaria** (*Xylaria*).

Xylaria poitei

If it is on wood, flame-shaped, flexible, and entirely gray and dusty, it is the anamorph stage of *Xylaria poitei* or *X. allantoidea*.

Xylaria

If it has an upper part that

- is pointed like a beak, and the whole structure is hollow, it is an immature **Texas star** (*Chorioactis geaster*).[A]

- is like a feathery brush, it is a **puzzlebrush** in *Anthina*.[ab]

- has perithecia, it is a **moonclub** (*Trichoderma leucopus* or *T. alutaceum*).

Otherwise, GO TO NARROW A PAGE 45

Trichoderma leucopus, pale moonclub

NARROW A

If there are two to a great many branches coming from a shared base, it is a **coral mushroom**—perhaps a **crown-tipped coral** (*Artomyces pyxidatus*), **crested coral** (*Clavulina*), *Phaeoclavulina*, *Ramaria*, *Ramariopsis*, or a **grand bonecoral** (*Sebacina schweinitzii*).

Artomyces pyxidatus, crown-tipped coral

If there are several narrow "tentacles" sprouting up from the same spot (but not branching from a shared base), it is a **spindle mushroom**. It is probably a *Clavulinopsis*, *Alloclavaria*, or *Clavaria*.

Clavaria, a spindle mushroom

If at least the top is strikingly bright yellow or orangish yellow, GO TO NARROW B PAGE 46

If it has a distinctly flattened or creased "head" on top, it is an earthtongue—perhaps a **fairyfan** (*Spathularia* or *Spathulariopsis*), **black earthtongue** (*Geoglossaceae*), or *Microglossum*.[A]

Spathularia, fairyfan

If the flesh is tough and leathery to corky, it is an immature polypore, yet to develop pores. GO TO NARROW C PAGE 46

Otherwise, it is a **club mushroom**—perhaps a **candlelight vigilclub** (*Multiclavula mucida*), **fairyhair club** (*Macrotyphula juncea*), *Clavaria*, or **giant club** (*Clavariadelphus*).

Clavariadelphus, giant club

NARROW B

If it has no distinguishable "head", or a long, thin one atop a short "stem," it is a **club mushroom** in *Clavaria* or *Clavulinopsis*.

Otherwise, it is an ascomycete. It is either an **orange earthtongue** (*Microglossum rufum*), a **yellow fairyfan** (*Spathularia flavida*), **swampbeacon** (*Mitrula elegans* group), or **irregular earthtongue** (*Neolecta*).[A]

Microglossum rufum, orange earthtongue

Neolecta irregularis, an irregular earthtongue

NARROW C

If it is brown and thinly velvety, it is an immature **tigereye** (*Coltricia*).

If it is smooth and bright red, orange, yellow, and white, from base to tip, it is an immature **reishi** (*Ganoderma*).

an immature *Ganoderma*, reishi

If it is strikingly bright orange, GO TO EFFUSED A PAGE 47

If it is formed from a great many distinct miniscule spherical beads, GO TO EFFUSED B PAGE 47

If it is distinctly hairy, fuzzy, or simply smoothish and powdery, GO TO EFFUSED C PAGE 48

Otherwise, GO TO EFFUSED E PAGE 49

EFFUSED A

If it is a thick coat of slime dripping off of cut or injured wood, it is **spring flux**, an inconsistent assemblage of many bacterial and fungal species.[abfx]

If it is hard, matte, on unburnt wood, and solid or brittle, it is a **glowing woodwart** (*Hypoxylon crocopeplum*).[a]

Hypoxylon crocopeplum, glowing woodwort

If it is separate or continuously fused lumps, on burned wood or burned soil, it is a **firecrust** (*Pyronema omphalodes*).[a]

Pyronema omphalodes, firecrust

If it is on a rotting polypore, it is *Hypomyces aurantius*.[a]

If it is mainly composed of branching, fibrous/silky rootlike cords, it is the rhizomorphs of a **flaming toothcrust** (*Hydnophlebia chrysorhiza*).[b]

Hydnophlebia chrysorhiza, flaming toothcrust

EFFUSED B

If it is on wood, and

- the beads are hard and black (possibly partly white-coated), they are **carbon beads**—perhaps *Rosellinia*, *Ruzenia*, or *Lasiosphaeria*.[a]

Rosellinia

> • the beads are soft and white, it is a **couscous crust**
> (*Aegerita candida*).[b]

Aegerita candida, couscous crust

EFFUSED C

If it can easily withstand poking or rubbing, and

- • it is on Chinese privet, has dark-purple edges, and is smoothish-powdery, it is an **emperor crust** (*Virgariella*).[a]

Virgariella, emperor crust

- • it is intense dark orange and thickly hairy, carpet-like, **GO TO EFFUSED D PAGE 49**

- • it is outdoors, eye-catchingly pink to lilac to lavender, and fuzzy, it is *Punctularia atropurpurascens* or *Hypochnella violacea*.[b]

Punctularia atropurpurascens

Otherwise, delicate or not, it is probably a **mold**. It is probably an ascomycete (*Hypomyces*, *Penicillium*, *Kretzschmaria*, *Trichoderma*, or one of a great many others)[a] or a zygomycete (perhaps *Phycomyces*, *Pilobolus*, *Rhizopus*, or *Spinellus*)[f]. It might also be a basidiomycete mold (*Botryobasidium*)[b] or non-mold crust (e.g., *Tomentella*)[b].

Pilobolus, a mold

Hypomyces aurantius, a mold

EFFUSED D

If it is growing on a rock or part of a tree exposed to sunlight, it is **orange rockhair** (*Trentepohlia*), a genus of algae.[x]

Trentepohlia, orange rockhair

If it is growing on wood or in a shady place or indoor walls, it is an **ozonium**[b], the ambitious mycelium of an **inkcap** species in *Coprinellus* section *Domestici*.

Coprinellus sect. *Domestici*, an inkcap

EFFUSED E

If it is hard, inflexible, on wood, and it is entirely black, black-and-white, or dark, dull brick- or wine-colored, and it is

- clearly composed of strips or radiating branches, it is a **carbonlip** (*Hysteriales* or *Gloniales*) or a **carbonstar** (*Glonium stellatum*), respectively.[a]

- not, it is a **carbon crust**—perhaps a **brittle cinder** (*Kretzschmaria*), Biscogniauxia, Diatrype, Camillea, Hypoxylon, or Jackrogersella.[a]

Kretzschmaria deusta, a brittle cinder

49

If it has some color and has perithecia, it is a **mooncrust**. It is probably **coral hell** (*Helminthosphaeria clavariarum*) on the base of a crested coral, a *Hypomyces* on another mushroom, *Nectriopsis violacea* on a slime mold in *Fuligo*, or a *Trichoderma*.[a]

Helminthosphaeria clavariarum, coral hell

If it mainly consists of branching, rootlike cords, and it is

- consistently bright yellow, it is a **yellow rootweb** (*Piloderma bicolor*).[b]

Xenasmatella vaga, warrior rootweb

- mostly pale yellow, with light brown near the middle and with white near the tips, it is a **warrior rootweb** (*Xenasmatella vaga*).[b]

- entirely black, with wide flattened cords wrapped around a large segment of wood, it is the *rhizomorphs* (mycelial cords) of a **honey mushroom** (*Armillaria*).[b]

Armillaria, honey mushroom, with rhizomorphs

- entirely white, it is probably the unidentifiable **mycelium** of another fungus.[ab]

Otherwise, it is probably a true crust—perhaps **ceramic parchment** (*Xylobolus frustulatus*), **wet rot** (*Coniophora*), a crust in *Hymenochaetaceae*, **phlebie** (*Phlebia*,

Xylobolus frustulatus, ceramic parchment

Byssomerulius), **giraffe spot** (*Peniophora albobadia*), or **cobalt crust** (*Terana coerulea*), but it is probably one of countless less eye-catching others.[b]

Terana coerulea,
cobalt crust

 If it is a rounded cluster of many flattened lobes or chambers, it is a **cauliflower mushroom** (*Sparassis*), **clambering rosette** (*Irpex rosettiformis*), or **bonfire cabbagehead** (*Daleomyces phillipsii*)[A].

Sparassis cripsa,
cauliflower
mushroom

If it is soft, white, easily discoloring pale pinkish or pale yellowish, and irregularly popcorn-shaped with a rumpled outer surface (perhaps silky at the base), it is a **shrimp-of-the-woods**, a honey mushroom (*Armillaria*) parasitized by an abortive pinkgill (*Entoloma abortivum*).

Daleomyces phillipsii,
bonfire cabbagehead

If it is distinctly barrel-shaped, and 1 cm tall or less, it is an unopened **bird's-nest mushroom.**
SEE CUP A PAGE 38

Armillaria parasitized
by *Entoloma abortivum,*
shrimp-of-the-woods

If it is growing on wood, GO TO GLOBULAR A PAGE 52

If it is growing on the ground, GO TO GLOBULAR C PAGE 53

GLOBULAR A

If it is white and extraordinarily fuzzy all around, it is the mold-like stage of a **powderpuff bracket** (*Ptychogaster albus*).[x]

If slicing it in half vertically reveals a layer of pores inside, it is a **veiled polypore** (*Cryptoporus volvatus*).[x]

Cryptoporus volvatus, veiled polypore, with section showing veil and pores

If it is black (or dark reddish) and hard, it is a **carbon cushion**.[A] SEE CUSHION A PAGE 54

If it is distinctly folded and/or has perithecia, it is a **mooncushion**[A] or the anamorph stage of *Xylaria poitei*[A]. SEE CUSHION E PAGE 57

Entonaema liquescens, sulfur gush, a mooncushion

If it is >2 cm wide, with a tough-to-hard consistency, it is a **conk** that has not formed pores yet. SEE PORE D PAGE 23

Otherwise, GO TO GLOBULAR C PAGE 53.

GLOBULAR B

If it is <1.5 cm wide and entirely bright yellow, it is a **fractal moldtruffle** (*Endogone pisiformis*).[F] SEE PAGE 33

Endogone pisiformis, fractal moldtruffle

If it was fully or nearly fully buried in soil, it is a **truffle** (it might be a *Tuber* or *Leucangium*)[A] or **false-truffle** (it might be a **pogie**, *Rhizopogon,* or a **pax truffle**, *Melanogaster*).

Otherwise, it is either a **puffball** (*Lycoperdaceae*) or an **earthball** (*Pisolithus* or *Scleroderma*).

Tuber, a truffle

GLOBULAR C

If it has several pointed, petal-like rays pointing out or down around the circumference, it is an **earthstar** (*Astraeus, Geastrum,* or *Myriostoma coliforme*) or a **star earthball** (*Scleroderma polyrhizum*).

If, upon being sliced in half vertically, it

- is gelatinous, it is a **stinkhorn egg** or a **trufflehorn** (it might be *Protubera* or *Phallogaster*).

- displays two small "pockets" with tiny gills, it is an **amanita** egg (*Amanita*).

Otherwise, it is a typical gasteroid
GO TO GLOBULAR B PAGE 52.

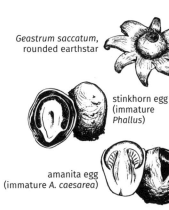
Geastrum saccatum,
rounded earthstar

stinkhorn egg
(immature
Phallus)

amanita egg
(immature *A. caesarea*)

If it is a mold-like tangle, spray, or thicket of grayish, delicate filaments, growing from

Spinellus fusiger,
mycena mold

- a mushroom, with pinheads on the filaments, it is **mycena mold** (*Spinellus fusiger*).[f]

- a mushroom, without pinheads on the filaments, it is **cloud mold** (*Syzygites megalocarpus*).[f]

Syzygites megalocarpus,
cloud mold

- dung, it is probably a **giant pinmold** (*Phycomyces*).[f]

If it is black or distinctly spongy, GO TO CUSHION A BELOW

Otherwise, GO TO CUSHION B PAGE 55

CUSHION A

If it is spongy and on or under a beech tree, it is an **aphid sootsponge** (*Scorias spongiosa*).[a]

Scorias spongiosa,
aphid sootsponge

If it is large (≥10 cm) and erupting from the trunk of a standing

- aspen or poplar tree, it is probably **corky bark disease**, caused by *Diplodia tumefaciens*.[a]

- cherry tree, it is probably a **black knot,** caused by *Apiosporina morbosa*.[a]

- birch or alder tree, and has a marbled rusty and yellowish-orange interior, it is **chaga** (the sclerotium of *Inonotus obliquus*).[b]

Inonotus obliquus, (anamorph), chaga

- tree otherwise, it is a burl or gall of another origin.[x]

Otherwise, it is a **carbon cushion**. It might be **dog's nose** (*Camarops petersii*)[A], **King Alfred's cake** (*Daldinia*)[A], **giant crampball** (*Annulohypoxylon thouarsianum*)[A], or a **woodwart** (*Hypoxylon* or *Jackrogersella*)[a].

Daldinia childiae, King Alfred's cake

CUSHION B

If it has a wood-like coloration, texture, and consistency, it is wood—either a gall of non-fungal origin (SEE PAGE 35) or a burl.[x]

Annulohypoxylon thouarsianum, giant crampball

If it is a gathering of more than a dozen small, tough-to-hard units, GO TO CUSHION C PAGE 56

If it is growing on the ground, tree roots, or the base of a stump, GO TO CUSHION D PAGE 56

If it is growing on dead wood, a trunk, or a conk, GO TO CUSHION E PAGE 57

CUSHION C

If they are orangish to reddish, and each one

- is bumpy like a blackberry, it is a **woodwart** in *Hypoxylon*.[a]

Peniophora rufa
rufous cushion

- is gently wrinkled like a flattened wad of gum, it is a **rufous cushion** (*Peniophora rufa*).[b]

- is rough and hard, it is an **orange-hobnail canker** (*Amphilogia gyrosa*).[a]

Amphilogia gyrosa,
orange-hobnail canker

If they are grayish and whitish, they are immature **dead man's fingers** (*Xylaria polymorpha*).[A]

CUSHION D

If it is growing in grass or other exposed areas in California, >2 cm wide, hard, and robust, it is a **swirling reishi** (*Ganoderma polychromum*) yet to develop pores.

Ganoderma polychromum,
swirling reishi

If it is mahogany with a pale edge, compressed, hugging the ground, and lumpy-smooth, it is a **doughnut fungus** (*Rhizina undulata*).[a]

Rhizina undulata,
doughnut fungus

CUSHION E

If it is light brown, dry, not solid, bumpy, messy, and powdery, it is the anamorph stage of an **irregular Midas-ear** (*Ionomidotis irregularis*).[A]

Ionomidotis irregularis, irregular Midas-ear (asexual state)

If it is rusty to wine to gray to light brown (not pink), and

- soft but firm, it is an immature (still veiled) **dog's nose** (*Camarops petersii*).[A]

Daldinia childiae, King Alfred's cake

- hard, it is a **King Alfred's cake** (*Daldinia*).[A]

If it is conspicuously dusty, it is the anamorph stage of *Xylaria poitei*.[A]

If it has perithecia, is not entirely solid, or forms ghastly yellowish solid "gum wads" affixed to the wood, it is a **mooncushion**. It is a *Trichoderma*, **sulphur gush** (*Entonaema liquescens*), or **sulphur blush** (*Thuemenella cubispora*).[A]

Entonaema liquescens, a sulphur gush

Otherwise, it is an immature polypore, yet to develop pores.

Thuemenella cubispora, sulphur blush

INDEX

Abortiporus biennis 19, 23
Acarospora 25
Acarospora contigua 7
Aceria 35
Aculops 35
Aegerita candida 48
Agaricus deserticola 42
Agrobacterium 35
Agrobacterium radiobacter 35
Akanthomyces 33
Alectoria sarmentosa 24
Aleuria aurantia 37, 38
Allodus podophylli 18
Allokermes 35
Allorhizobium 35
Amanita 17, 42, 53
Amanita muscaria 17
Amphibolips 35
Amphilogia gyrosa 56
Amylocystis lapponica 23
Annulohypoxylon
   thouarsianum 55
Anthina 44
Antrodia 19
Aphroditeola olida 15

Apiosporina morbosa 11,
   34, 54
Arcyria 13
Armillaria 50–51
Arrhenia retiruga 40
Artomyces pyxidatus 45
Ascocoryne 30–32
Ascomycota 4
Aseroe rubra 28
Asterophora lycoperdoides 40
Astraeus 53
Aureoboletus 20–21
Aureoboletus mirabilis 21
Auricularia 31
Austroboletus subflavidus 20

Badhamia 13
Baorangia 22
Basidiomycota 4
Battarrea phalloides 40
Beauveria 33
Besbicus mirabilis 35
Biscogniauxia 49
Bjerkandera 24
Boletellus ananas 21
Boletellus chrysenteroides 21
Boletinellus merulioides 21
Boletopsis 22

Boletus 22
Bondarzewia berkeleyi 23
Botryobasidium 48
Buchwaldoboletus 21
Bulgaria inquinans 31
Butyriboletus 22
Byssomerulius 40, 51

Callirhytis 35
Caloboletus 22
Calocera 32
Calostoma 42
Calvatia cyathiformis 38
Calycina citrina 38
Camarops petersii 55, 57
Camillea 49
Cantharellus 15
Cantharellus lateritius 41
Ceratiomyxa fruticulosa 12
CeriOporus 22
Ceriporia 19
Cerrena unicolor 24
Chalciporus 22
Chlorociboria 38
Chlorophyllum 17
Chlorophyllum molybdites 17
Chondrostereum
   purpureum 40

Chorioactis geaster 36, 44
Chrysothrix 259
Cladonia 25
Clathrus archeri 28
Clathrus columnatus 28
Clavaria 45
Clavariadelphus 45
Claviceps 36
Claviceps purpurea 36
Clavicipitaceae 36
Clavulina 45
Climacocystis borealis 23
Clitopilus prunulus 14
Collema 30
Coltricia 16, 22, 46
Coltricia montagnei var
   greenei 16
Coniophora 50
Conopholis 8
Coprinellus sect Domestici 49
Cordyceps 11, 33
Cordyceps tenuipes 11
Cortinarius 17
Cotylidia diaphana 41
Craterellus 15, 41
Crepidotus 14–15
Cronartium 34
Cryptoporus volvatus 52

Cudonia spathulata 43
Cudoniella acicularis 43
Cudoniella clavus 43
Cyathus striatus 38
Cymatoderma caperatum 41

Dacrymyces 29, 31–32
Dacrymyces chrysospermus 29
Dacrymyces corticioides 31
Dacryopinax elegans 31
Daedaleopsis confragosa 14
Daldinia 41, 55, 57
Daleomyces phillipsii 51
Diastrophus 35
Diatrype 49
Diplodia tumefaciens 54
Disholcaspis 35
Ductifera pululahuana 30

Echinodothis 36
Endogone pisiformis 33, 52
Engyodontium 33
Entoloma stricta 17
Entolomataceae 17
Entonaema liquescens 52, 57
Eocronartium muscicola 32
Epichloe 36
Erynia 33
Euura 35

Evernia 25
Exidia glandulosa 30
Exidia recisa 30–31
Exobasidium 35
Exsudoporus frostii 20

Fistulina americana 18
Flavoparmelia 25
Fomes 23
Fomitiporia 19
Fomitopsis 14, 23
Fomitopsis quercina 14
Fulvifomes 23
Fuscoporia gilva 24

Ganoderma 6, 22–23, 46, 56
Ganoderma applanatum 6
Ganoderma polychromum 56
Geastrum 53
Geastrum saccatum 53
Geoglossaceae 45
Gibellula 33
Gloeomucro 32
Gloeophyllum sepiarium 14
Gloniales 49
Glonium stellatum 49
Gomphidius 17
Gomphus 15
Grifola frondosa 23

Guepinia 32
Guepiniopsis 32
Gymnoconia 18, 34
Gymnoconia peckiana 18
Gymnosporangium 27, 29, 32
Gymnosporangium juniperi-
    virginianae 34
Gyromitra 40, 44
Gyroporus 20

Helminthosphaeria
    clavariarum 50
Helvella 38, 40, 44
Helvella macropus 38
Helvellosebacina
    concrescens 30
Hemileccinum 20
Henningsomyces 18
Herpothallon rubrocinctum 25
Heyderia abietis 43
Hohenbuehelia 15
Holwaya 43
Hortiboletus 21
Humaria 38
Hydnophlebia chrysorhiza
    27, 47
Hydnoporia tabacina 39
Hymenochaetaceae 24, 50

Hymenochaete rubiginosa 39
Hypholoma 17
Hypochnella violacea 48
Hypomyces 14, 40, 42, 47,
    48, 50
Hypoxylon 37, 47, 49, 55–56
Hypoxylon crocopeplum 37,
    47, 56
Hysteriales 49

Imleria badia 21
Inonotus hispidus 24
Inonotus obliquus 55
Inonotus rickii 12
Ionomidotis irregularis 57
Irpex rosettiformis 51

Jackrogersella 49, 55

Kretzschmaria deusta 49
Kusaghiporia persicina 23

Lactarius 17
Lactifluus 17
Laetiporus 23–24
Laetiporus cincinnatus 23
Laetisaria fuciformis 36
Laetisaria roseipellis 36
Lanmaoa 21–22
Lasiosphaeria 47

Leccinum 20
Lentinellus 15
Lentinus 22, 42
Lentinus tigrinus 42
Leotia 32
Lepista 17
Lepraria 25
Leptogium 30
Leptoporus mollis 23
Letharia 25
Leucangium 53
Leucocoprinus 17
Lobaria 25
Lycogala 12
Lycoperdaceae 42, 53
Lysurus 28

Macrocystidia 16
Macrolepiota 17
Macrotyphula juncea 45
Massospora 33
Melanogaster 53
Melanophyllum 16
Melanopsichium 35
Mensularia radiata 24
Meripilus sumstinei 23
Merismodes 18
Metarhizium 33

Microglossum 45–46
Microglossum rufum 46
Microporellus dealbatus 22
Mitrula elegans 43, 46
Monotropa 8
Morchella 43
Morchella americana 43
Mordwilkoja 35
Multiclavula mucida 45
Mutinus 28
Mycosarcoma maydis 36
Myriostoma coliforme 53
Myxarium nucleatum 30
Myxomycete 7

Naematelia aurantia 29
Naematelia encephala 30
Nanokermes 35
Nectriopsis violacea 50
Neoboletus 21
Neobulgaria 30–32
Neofavolus alveolaris 24
Neolasioptera 35
Neolecta irregularis 46
Nidulariaceae 38–39
Normandina pulchella 25
Nostoc commune 29

Onnia 23

Ophiocordyceps 33

Pachypsylla 35
Panaeolus 17
Panus lecomtei 2
Paraisaria 33
Peltigera 25
Peniophora albobadia 5, 51
Peniophora rufa 56
Phaeoclavulina 45
Phaeolepiota aurea 17
Phaeolus schweinitzii 23
Phaeotremella 30
Phallus 28, 53
Phallus rubicundus 28
Phellinus 23
Phlebia 40, 50
Phlebia tremellosa 40
Phleogena faginea 13, 42
Pholiota nubigena 42
Phragmidium 34
Phycomyces 48, 54
Phylloporia amplectens 24
Phyllotopsis nidulans 14
Phylloxera 35
Phylloxerid 35
Physalacria inflata 41
Picipes 22

Pileolaria brevipes 34
Piloderma bicolor 50
Placidium arboreum 25
Pleurocordyceps 33
Pleurocybella porrigens 15
Pleuroflammula flammea 15
Pleurotus 15
Pluteaceae 17
Podaxis 42
Podoscypha 41
Polyozellus 15
Polyporus umbellatus 22
Polystepha 35
Poronia 40
Porotheleum fimbriatum 18
Porpidia 25
Postia 23
Prospodium transformans 34
Protubera 53
Pseudoboletus parasiticus 21
Pseudocolus fusiformis 28
Pseudomerulius curtisii 14
Pseudoplectania nigrella 31
Psora 7, 25
Psora pseudorussellii 7
Ptychogaster albus 52
Puccinia 2, 18, 27, 34
Pulveroboletus ravenellii 21

Punctularia atropurpurascens 48
Punctularia strigosozonata 39
Purpureocillium 33
Pycnoporellus 19
Pycnoporellus alboluteus 19
Pyronema omphalodes 47

Ramalina 24–25
Ramalina menziesii 24
Ramaria 45
Ramariopsis 45
Rectipilus 18
Retiboletus 22
Rhizina undulata 56
Rhizopogon 53
Rhodotus 15
Rhopalomyia 35
Rosellinia 47
Rubroboletus 21
Rubroboletus eastwoodiae 21
Russula 17, 40, 42
Russula xerampelina 17
Ruzenia 47

Schizophyllum commune 13
Scleroderma polyrhizum 36, 53
Scorias spongiosa 34, 54

Scutigeraceae 22
Sebacina incrustans 34
Sebacina schweinitzii 45
Sebacina sparassoidea 32
Simocybe haustellaris 15
Sorosphaerula 35
Sparassis 37, 51
Sparassis crispa 37
Spathularia 45–46
Spathularia flavida 46
Spathulariopsis 45
Spinellus 48, 54
Spinellus fusiger 54
Stemonitis 7, 13
Stemonitopsis 13
Stereopsis hiscens 41
Stereopsis humphreyi 41
Stilbella fimetaria 2
Strobilomyces 21
Suillellus 21
Suillus 2, 20–21
Suillus ampliporus 20
Suillus cavipes 2
Suillus luteus 20
Synchytrium geranii 33
Synchytrium papillatum 33
Syzygites megalocarpus 54
Syzygospora mycetophila 30

Taphrina 35
Tapinella atrotomentosa 16
Tapinella panuoides 15
Terana coerulea 51
Testicularia 36
Tetraneura 35
Thelephora 40–41
Thuemenella cubispora 57
Tolypocladium 42
Trametes 13, 24, 39
Trametes betulina 13
Trametes conchifer 39
Tremella fuciformis 30
Tremella mesenterica 29
Trentepohlia 49
Trichaptum 24
Trichoderma 42, 44, 50, 57
Trichoderma alutaceum 42, 44
Trichoderma leucopus 42, 44
Trioza 35
Tuber 53
Tulasnella aurantiaca 29
Turbinellus 15
Tylopilus 22
Tyromyces 23

Umbilicaria 25
Urnula craterium 38

Uromyces ari-triphylli 18
Usnea 24
Ustilaginales 36
Ustilago 36
Ustilago nuda 36

Verpa 43
Vibrissea truncorum 43
Virgariella 48
Viridothelium virens 25

Xanthomendoza 25
Xenasmatella vaga 50
Xerocomellus 21
Xerocomus 21
Xylaria 11, 44, 52, 56–57
Xylaria flabelliformis 11
Xylaria poitei 52, 57
Xylaria polymorpha 56
Xylobolus frustulatus 50
Xylodon 19
Xylodon raduloides 19

Other books in the pocket-size *Finder* series:

FOR US AND CANADA, EAST OF THE ROCKIES

Berry Finder native plants with fleshy fruits
Bird Finder frequently seen birds
Bird Nest Finder aboveground nests
Fern Finder native ferns of the Midwest and Northeast
Flower Finder spring wildflowers and flower families
Life on Intertidal Rocks organisms of the North Atlantic Coast
Scat Finder mammal scat
Track Finder mammal tracks and footprints
Tree Finder native and common introduced trees
Winter Tree Finder leafless winter trees
Winter Weed Finder dry plants in winter

FOR THE PACIFIC COAST

Pacific Coast Bird Finder frequently seen birds
Pacific Coast Fish Finder marine fish of the Pacific Coast
Pacific Coast Mammal Finder mammals, their tracks, skulls, and other signs

FOR THE PACIFIC COAST *(continued)*

Pacific Coast Tree Finder native trees, from Sitka to San Diego
Pacific Intertidal Life organisms of the Pacific Coast
Redwood Region Flower Finder wildflowers of the coastal fog belt of CA

FOR ROCKY MOUNTAIN AND DESERT STATES

Desert Tree Finder desert trees of CA, AZ, and NM
Rocky Mountain Flower Finder wildflowers below tree line
Rocky Mountain Mammal Finder mammals, their tracks, skulls, and other signs
Rocky Mountain Tree Finder native Rocky Mountain trees

FOR STARGAZERS

Constellation Finder patterns in the night sky and star stories

NATURE STUDY GUIDES are published by AdventureKEEN, 2204 1st Ave. S., Suite 102, Birmingham, AL 35233; 800-678-7006; naturestudy.com. See shop.adventurewithkeen.com for our full line of nature and outdoor activity guides by ADVENTURE PUBLICATIONS, MENASHA RIDGE PRESS, and WILDERNESS PRESS, including many guides for birding, wildflowers, rocks, and trees, plus regional and national parks, hiking, camping, backpacking, and more.